Animal Mysteries

Green Spaces

Forests, Grasslands, & Jungles

*Engaging Mysteries
for Younger Minds to Solve*

Ages 5+

Adam C Laningham

Animal Mysteries

A Bright Child Books LLC Publication

All rights reserved. No part of this book may be reproduced in any manner, except for the use within an individual classroom or individual parent without written permission from the publisher. For more information, please contact: adam@brightchildbooks.com

BRIGHT CHILD BOOKS, LLC
BrightChildAZ.com
BOOKS, PUBLISHING, & MORE

Printed in USA

Design and Cover Art by Adam C. Laningham

Why I Wrote This Book

For more than four decades, **Nathan Levy's Stories With Holes** have been used by teachers and parents around the world to engage thousands of children in creative, critical, and rigorous thinking. These classic stories challenge students to use reasoning, cultural knowledge, and problem-solving skills in fun and unexpected ways.

However, because of the level of reasoning and cultural literacy they require, the original Stories With Holes have often been best suited for older or more advanced students. Over the years, countless teachers and parents have asked if we could create a version designed especially for younger learners.

In 2026 we released **Stories With Holes Junior** which is great for first and second graders, but we still wanted something for even younger children or those new to the inquiry process and needed a bit more foundation.

After much thought, we realized the perfect topic for a mystery-style guessing game that younger children already love and understand — **animals!**

This new collection brings the engaging, riddle-like format of *Stories With Holes* to early elementary students through fun, animal-themed mysteries. The familiar subject matter makes the challenges accessible, while still building the problem-solving, reasoning, and creative-thinking skills children will need for the original *Stories With Holes* as they grow.

How to Use These Animal Mystery Activities

First of all, please feel free to modify anything to make this book work for you and your classroom or child.

Teacher Tips:

These mysteries are flexible tools for literacy, science, and group problem-solving. They work well as warm-ups, centers, or integrated lessons.

How to Use:

- Explain to the class that these are all animals that are likely to be found _____.
- Whole Class Morning Starter or Brain Break Activity. The teacher reads the clues out loud slowly, giving time for children to think in between each clue.
  - You can have children raise their hands to answer out loud each time a clue is given
  - The Children can write down their guesses on a piece of paper, then adjust their guess based on each clue and see how many times they change their guess

Animal Mysteries

- Team Detectives: Same as the whole class activity but divide the class into groups. Each team discusses clues and presents their answer with evidence.
- Cross-Curricular Extensions of Links:
    - Science: Connect the animal to its habitat, diet, or adaptations. There are several versions of the series to tie in two different animal habitats!
    - Language Arts: Have students write their own mystery stories in the same format. maybe they can go beyond animals and pick other things?
    - Drama: Have students act out animal movements or sounds to see if the other team can guess their animal.
- Difficulty Scaling: Use simpler sets for younger grades and challenging sets for advanced thinkers.
- Assessment Tool: Observe how students reason—do they jump to the easy clue, or infer from harder ones? Have them reflect on their problem-solving abilities.
- **Encourage children to reflect on the clues they did not know so they learn about these fascinating creatures!**
- Don't tell the children the titles of the mystery until afterward.

Benefits:

- Encourages teamwork and discussion.
- Strengthens vocabulary and comprehension.
- Integrates science and literacy in a fun way.
- Builds confidence as students "solve" mysteries together.
- Increases problem-solving abilities
- **Encourages inquiry and the love and mystery of learning!**

Parent Tips:

These animal mysteries are designed to spark curiosity and playful learning. Parents can use them to build listening, reasoning, and imagination skills while bonding with their child.

How to Use:

- Bedtime Detective: Read one mystery aloud before bed. Pause after each clue to let your child guess.
- Clue Countdown: Start with the hardest clue. Encourage guesses after each one, refining as the clues get easier.
- Family Challenge: Take turns being the "mystery reader." Kids love stumping parents too!

Animal Mysteries

- Movement Play: Act out the animal once solved—crawl like a crab, hop like a rabbit, roar like a lion.
- Creative Extension: After solving, ask your child to invent one new clue or fact about the animal.
- Encourage your child to research the animals further or research other animals in that environment.
- Encourage children to reflect on the clues they did not know so they learn about these fascinating creatures!
- Don't tell the children the titles of the mystery until afterward.

Benefits:

- Builds listening and inference skills.
- Encourages curiosity about nature and habitats.
- Strengthens family connection through shared play.
- Supports creativity and memory recall.
- **Encourages inquiry and the love and mystery of learning!**

Animal Mysteries Green Spaces

I am an animal that can be found in the forest…

Animal Mysteries

1. The Echoing Drummer

1. I cling to tree trunks with strong claws.
2. I search for hidden insects under bark.
3. My red crest flashes as I peck.
4. I make a sound like a drum, though I carry no sticks.
5. My tapping echoes through the forest.
6. My beak is sharp enough to carve into wood.

Answer: Woodpecker

2. The Silent Stalker

1. My eyes glow like lanterns in the night.
2. I fly without making a sound.
3. I hunt small creatures with sharp talons.
4. My head can turn almost all the way around.
5. I hoot softly from high branches.

Answer: Owl

Animal Mysteries

3. The Antlered King

1. I am known for my size and strength.
2. I move quietly through the forest.
3. My antlers grow larger each year, branching like trees.
4. I graze on leaves and twigs.
5. My hooves leave deep prints in the soil.

Answer: Moose

4. The Masked Bandit

1. I sneak through campsites at night, searching for food.
2. I'm known as the trickster of the forest.
3. My paws are nimble, almost like hands.
4. I wash my meals before eating.
5. My fur is gray, with a striped tail.
6. My black mask makes me look like a thief.

Answer: Raccoon

5. The Bushy-Tailed Acrobat

1. I have sharp teeth.
2. I leap from branch to branch with ease.
3. I chatter loudly when I'm upset.
4. I hide food in secret spots for winter.
5. My tail is long and fluffy.
6. I'm the nut collector of the forest.

Answer: Squirrel

6. The Tiny Builder

1. I carry loads many times my size.
2. I march in lines across the forest floor.
3. I dig tunnels underground.
4. I work with thousands of others.
5. I bite if disturbed.
6. I'm the smallest but busiest worker in the forest.

Answer: Ant

Animal Mysteries

7. The Shaggy Climber

1. I lumber slowly, but climb trees with ease.
2. My claws are long and curved.
3. I eat berries, fish, and honey.
4. My growl shakes the forest.
5. My thick fur keeps me warm.
6. I'm the largest predator in the forest.

Answer: Bear

8. The Whispering Jumper

1. I move only at night, staying hidden.
2. I eat insects and small creatures.
3. My long legs help me leap silently.
4. My eyes shine when caught in light.
5. My croak echoes near ponds.
6. I'm the leaping singer of the forest.

Answer: Frog

9. The Slippery Trickster

1. I bask in sunlight to warm my body.
2. I eat rodents and other small animals
3. My tongue flicks to taste the air.
4. My scales shimmer in green or brown.
5. I coil tightly when threatened.
6. I'm the slithering reptile of the forest.

Answer: Snake

10. The Antlered Wanderer

1. I move silently through the undergrowth.
2. I graze on leaves and twigs.
3. My babies have spots to hide for danger.
4. I may even live in your neighborhood.
5. My antlers grow larger each year.

Answer: Deer

11. The Busy Builder

1. I eat leaves and twigs
2. I have a brown watertight furcoat
3. I gnaw wood with sharp teeth.
4. My tail is flat like a paddle.
5. I build dams across streams.
6. I'm known as the forest engineer.

Answer: Beaver

12. The Jumper

1. My coat blends with dappled light.
2. I stalk prey silently.
3. My claws grip tightly.
4. My roar is softer than a lion's.
5. I'm the leopard of the forest.

Answer: Bobcat

13. The Shy Burrower

1. I dig tunnels under forest soil.
2. My body is long and furry.
3. I eat insects and small reptiles.
4. I rarely come out in daylight.
5. My fur is brown and gray but my face is black and white.

Answer: Badger

14. The Red Flash

1. I eat seeds and berries.
2. My song is clear and cheerful.
3. I perch high in trees.
4. Many people put birdseed out for me to eat.
5. My feathers are bright scarlet.

Answer: Cardinal

15. The Thorny Sleeper

1. I shuffle slowly through the leaves.
2. I eat insects
3. I usually come out at night.
4. I curl into a ball when scared.
5. My body is covered in sharp spines.

Answer: Hedgehog

16. The Silent Stalker

1. I am known for my intelligence.
2. I can move quietly through the undergrowth.
3. My ears perk at every sound.
4. I hunt rabbits and mice.
5. My coat is reddish brown.

Answer: Fox

15. The Silent Shadow

1. My body is sleek and muscular.
2. I stalk prey silently through trees.
3. I hunt large prey.
4. My tail helps me balance when leaping.
5. My cry is a piercing scream.

Answer: Mountain Lion

16. The Spiny Climber

1. I shuffle slowly through the undergrowth.
2. I can climb trees.
3. I eat bark and leaves.
4. Predators learn to avoid me.
5. My body is covered in sharp quills.

Answer: Porcupine

17. The Secretive Hunter

1. My body is long and slender.
2. I move silently through branches.
3. You may have one of my cousins as a pet.
4. My fur is sleek and dark.
5. I prey on squirrels and small birds.

Answer: Pine Marten/weasel

18. The Silent Tracker

1. My paws leave prints in soft soil.
2. I move quietly, sniffing the air.
3. I hunt large animals with the help of my family.
4. My fur is gray or brown.
5. I howl to call my pack.

Answer: Wolf

19. The Forest Drummer

1. I eat seeds and berries.
2. My wings flash bronze and green but I rarely fly..
3. I strut proudly on the forest floor.
4. My call is a deep "thump-thump."
5. I remind people of autumn.

Answer: Wild Turkey

20. The Hidden Jumper

1. My skin is smooth and moist.
2. I mostly come out at night.
3. I live near ponds and streams.
4. I leap through the damp leaves.
5. I croak softly near streams.

Answer: Toad**

Animal Mysteries

Animal Mysteries Green Spaces

I am an animal that can be found in the grasslands…

Animal Mysteries

1. The Striped Runner

1. I graze in herds across open plains.
2. My hooves thunder when we gallop together.
3. No two of us look exactly alike.
4. My coat is patterned like a barcode.
5. I look like a horse but wear stripes.

Answer: Zebra

2. The Shaggy Grazer

1. I graze on grass in herds.
2. My thick fur keeps me warm in winter.
3. My shoulders are humped with muscle.
4. I once roamed the Great Plains in millions.
5. I have horns and am taller than a person.

Answer: Bison (American Buffalo)

3. The Horned Fighter

1. I graze on grass in herds.
2. My body is heavy and covered in thick skin.
3. I charge when threatened, head lowered.
4. My horn is sharp and strong.
5. I'm the armored grazer of the savanna.

Answer: Rhinoceros

4. The Tall Watcher

1. I walk slowly but cover great distances.
2. I browse leaves others cannot reach.
3. My spots help me blend with branches.
4. My neck stretches higher than most trees.
5. I'm the tallest animal on Earth.

Answer: Giraffe

5. The Mane King

1. I live in groups.
2. I hunt mostly at night.
3. My roar can be heard miles away.
4. My mane makes me look larger.
5. I'm the king of the grasslands.

Answer: Lion

5. The Burrowing Barkers

1. I live in family groups.
2. I eat grass and seeds.
3. I borrow tunnels under the Earth.
4. I stand upright to watch for danger like hawks and coyotes.
5. My bark sounds like a squeaky toy.

Answer: Prairie Dog

6. The Swift Sprinter

1. I chase prey across open plains.
2. Even though I am not a dog my call sounds like a bark.
3. My body is built for speed.
4. My spots are small and round.
5. I can run faster than any other land animal.

Answer: Cheetah

7. The Long-Horned Wanderer

1. I graze calmly in herds.
2. My body is strong and muscular.
3. I'm a symbol of the western plains.
4. My horns stretch wide like giant branches.
5. I'm the symbol of Texas.

Answer: Longhorn Cattle

Animal Mysteries

9. The Tiny Jumper

1. My body is small.
2. My legs are strong.
3. I leap quickly through tall grasses.
4. I chew plants with sharp jaws.
5. I chirp loudly in summer.

Answer: Grasshopper

10. The Silent Stalker

1. My coat blends with tall grass.
2. I crouch low before leaping.
3. My claws grip prey tightly.
4. My roar is softer than a lion's.
5. I'm the spotted cat of the plains.

Answer: Leopard

Animal Mysteries
Green Spaces

I am an animal that can be found in the jungle…

Animal Mysteries

1. The Striped Stalker

1. I move silently through the jungle and forest.
2. My roar shakes the jungle.
3. I hunt alone, not in packs.
4. My orange coat blends with tall grass.
5. My stripes are like fingerprints—no two alike.
6. I'm the largest cat in the world.

Answer: Tiger

2. The Long-Armed Acrobat

1. I swing from branch to branch with ease.
2. My arms are longer than my legs.
3. I rarely come down to the ground.
4. I call loudly, echoing through the trees.
5. I eat fruit high in the canopy.
6. I am not a monkey.

Answer: Gibbon

3. The Silent Slitherer

1. I bask in sunlight to warm my body.
2. My tongue flicks to taste the air.
3. Some of my kind squeeze prey, others bite.
4. I move without legs, gliding through leaves.
5. I coil tightly when threatened.

Answer: Snake

4. The Trumpeting Giant

1. I live in groups led by females.
2. My footsteps shake the ground.
3. I leave wide trails through the forest.
4. My ears flap like giant fans.
5. I use my trunk to grab food and spray water.

Answer: Elephant

5. The Colorful Climber

1. I eat insects with a quick tongue.
2. I cling to branches with sticky toes.
3. My body is small but my colors are bright.
4. My skin can warn predators of danger.
5. I croak loudly in the rain.

Answer: Poison Dart Frog

6. The Loud Howler

1. I eat leaves, fruit, and flowers.
2. I live high in the canopy.
3. My tail grips branches like a hand.
4. My voice can be heard miles away.
5. I travel in noisy groups.
6. I'm the monkey that howls at dawn.

Answer: Howler Monkey

7. The Stealthy Predator

1. I stalk prey silently, waiting to pounce.
2. My coat blends with dappled light.
3. My jaws are powerful enough to crush bone.
4. I climb trees to drag prey out of reach.
5. My spots are rosettes, not stripes.

Answer: Jaguar

8. The Slow Climber

1. My claws are long and curved.
2. I eat leaves and sleep for hours.
3. I move so slowly that moss grows on my fur.
4. I spend most of my life hanging upside down.
5. Predators often miss me because I barely move.
6. I'm the slowest mammal in the jungle.

Answer: Sloth

9. The Bright-Beaked Bird

1. I live in pairs high in the canopy.
2. I eat fruit, tossing pieces with my bill.
3. My wings are short but strong.
4. My call is harsh and loud.
5. My beak is huge and colorful.
6. I'm the rainbow-beaked bird of the jungle.

Answer: Toucan

10. The Giant Leaper

1. My body is powerful, built for jumping.
2. I move mostly at night, hunting silently.
3. My eyes glow in the dark.
4. My tail helps me balance when leaping.
5. I eat smaller animals in the forest.
6. I'm the spotted cat that leaps through the jungle.

Answer: Leopard

Animal Mysteries

Animal Mysteries

About The Author

Adam C. Laningham brings more than two decades of experience in public education and a deep passion for supporting gifted and twice-exceptional learners. Recognized as *Arizona Gifted Teacher of the Year*, Adam has taught across multiple grade levels, developed and led innovative gifted programs, and managed district-wide initiatives serving over 6,000 gifted students.

As founder of Bright Child Books, Adam is the author of several works dedicated to nurturing the potential of advanced learners. He also established The Gifted Collective, a thriving learning center that supports gifted and 2e children and their families.

Adam's leadership extends nationally and internationally. He served six years on the board of Supporting the Emotional Needs of the Gifted (SENG), including serving as President, and contributes his expertise as a member of the Arizona Association for Gifted and Talented (AAGT) Advisory Council, a founding member of Callisto (supporting gifted foster youth), and an advisor for CogAT Riverside Insights. Most recently, he founded the nonprofit Growing Intellect & Fostering Talent (GIFT).

Recognized as an engaging speaker and trusted consultant, Adam is committed to creating opportunities, resources, and communities that empower gifted learners to thrive—academically, socially, and emotionally.

adam@brightchildbooks.com

Animal Mysteries

Further Resources
Get Younger Kids Thinking!

PreK - 1st Grade
Animal Mysteries

Grades 2-3
Stories With Holes Junior

Gradws 3+
Stories With Holes

www.BrightChildBooks.com
Books, Training, & More

Animal Mysteries

Further Resources
Get Your Kids Thinking!

www.BrightChildBooks.com
Books, Training, & More

Animal Mysteries

Further Resources
Get Your Kids Thinking!

Real world math that is fun, meaningful, and interesting!

www.BrightChildBooks.com
Books, Training, & More

Further Resources
Gifted, Twice-exceptional, & Trauma Support

www.BrightChildBooks.com
Books, Training, & More

Further Resources
Get Your Kids Thinking!

www.BrightChildBooks.com
Books, Training, & More

Thank you for purchasing one of our resources! We are continuously working on improving our titles, so we would love your input. If you see anything that needs to be adjusted or have any recommendations, please email us.

adam@BrightChildBooks.com

Also, we would love it if you could take a minute and leave us a positive review on Amazon. Every little bit helps small publishers and authors get their work out there!

Made in the USA
Coppell, TX
21 February 2026

71909600R00024